Permaculture: A Step by Step Guide for Home Permaculture

How to Become an Expert in Home Gardening Techniques

By: Wesley Dios

I0493965

PUBLISHERS NOTES

Disclaimer

This publication is intended to provide helpful and informative material. It is not intended to diagnose, treat, cure, or prevent any health problem or condition, nor is intended to replace the advice of a physician. No action should be taken solely on the contents of this book. Always consult your physician or qualified health-care professional on any matters regarding your health and before adopting any suggestions in this book or drawing inferences from it.

The author and publisher specifically disclaim all responsibility for any liability, loss or risk, personal or otherwise, which is incurred as a consequence, directly or indirectly, from the use or application of any contents of this book.

Any and all product names referenced within this book are the trademarks of their respective owners. None of these owners have sponsored, authorized, endorsed, or approved this book.

Always read all information provided by the manufacturers' product labels before using their products. The author and publisher are not responsible for claims made by manufacturers.

Paperback Edition

Manufactured in the United States of America

DEDICATION

This book is dedicated to my mentor Jason Murdock. He encouraged me to follow never be afraid to study other disciplines. It would only make me a more rounded person.

TABLE OF CONTENTS

Chapter 1- Making Use Of Animals Without Livestock

Permaculture is a way of dancing with nature. If we push and pull nature harshly to try and control her — as is often the case in our modern world — then the dance will quickly break down, and our partner may even push or pull in the opposite direction, knocking us down. However, if we pay attention to her cues — subtly leading her at times, following her lead in others — we can come together to produce a beautiful work of art. In the dance that is permaculture neither nature nor people can be as effective by themselves as they can be by working together.

It is easy to mistake permaculture for a form of organic gardening. Indeed, one of the descriptions I have heard of permaculture is that it is a revolution disguised as organic gardening. But where it may divert from gardening is that permaculture more consciously

Permaculture

seeks to link together many different forms of life. Plants play a major role — but so do bacteria, fungi, and animals. In fact, just as in nature, permaculture systems are at their most vibrant and productive when they integrate all different kingdoms of life into a complex whole.

In this chapter, we're going to zoom in on the role of animals. Now, the classic example of incorporating animals into a permaculture, or even gardening/farming operation, is to turn to livestock. Chickens, pigs, goats, sheep, cattle — all of them, and more, can contribute to soil fertility as well as help increase pasture growth and soil fertility through integrated management practices. However, I live on a 1.3 acre lot so getting large livestock is out of the question for me, and the demands on my time from work and family presently preclude even getting small livestock without them quickly becoming a burden.

Yet, I'm not giving up on integrating animals into my operation; quite the contrary. I encourage earthworms to aerate garden beds and turn organic mulch into rich worm castings. By incorporating ornamental flowers into my operation, I attract beneficial insects to both pollinate vegetables and prey on "pests". These should be relatively obvious to any somewhat experienced gardener. There is one other species that I have in abundance that I have not sufficiently incorporated into my operation — birds.

My house is situated in an open area ringed by trees, and numerous forest stands, both large and small, populate the surrounding area. All of this provides a lot of bird habitat. Ever since we moved into this house, we have put a bird feeder outside our living room window, and have taken great pleasure in watching the birds come and visit. We get blue jays, red-winged blackbirds, grackles, sparrows, starlings, hairy woodpeckers, red-bellied woodpeckers, juncos, goldfinches, titmouses, cardinals, nuthatches, chickadees, doves, and many other varieties too numerous to list here. They are especially busy in the winter, as our feeder provides one of the few reliable food supplies available for them.

Yet, even as I have attracted all of these birds into a space that I more closely share with them, I have failed to really integrate them into my permaculture systems. These birds can provide valuable contributions to my systems — especially in terms of contributing rich manure. While cardinal, blackbird and blue jay manure certainly isn't on the same level as chicken manure — especially in terms of gathering, composting and applying it in a targeted fashion — it isn't something to dismiss either. I would bet that if I could help the birds better "target" their droppings closer to my food production; it could prove a significant resource.

In that vein, I'm going to incorporate bushes favored by birds as a food source — especially in winter — as part of my fruit tree, bramble and bush guilds around my front yard swale. One I like in particular is winterberry, which can be found both as a tree and a bush. As it loses its leaves in winter, its bright red berries remain in large, tight clusters, providing a striking display of color in the midst of what is often a more dreary season in the northeast.

Sumacs, elderberry, arrow wood and barberry are examples of other bird-friendly plants that can provide both food and shelter. In exchange, the birds provide us with a valuable "waste" product and, perhaps more important, hours of enjoyment. When designing your systems, don't limit yourself to plants that are primarily beneficial to people only. By consciously including the needs of wildlife into our permaculture systems, we will undoubtedly increase yields through greater biodiversity while reducing the outside inputs we introduce; we will also reduce our workload in the process.

CHAPTER 2- PERMACULTURE PLANNING -SEED ORDERS

One of the activities that I most look forward to each year as I come to the "back side" of winter is ordering seeds for the next year's gardens. One of the biggest dangers of this activity however, is that it becomes like a gourmand going to a delicious all-you-can-eat buffet — my eyes become bigger than the amount of space and time I have to get everything done!

This year I concentrated my order on two sources: Johnny's Selected Seeds and Burpee Seeds. I've always found Johnny's to be one of the best sources of seed around — they have consistently high levels of germination for a year after the purchase date and they stock a wide range of varieties that include many heirlooms. Mother Earth News recently agreed with this sentiment, as they ranked Johnny's the #1 source of seeds for the home gardener. However, as good as Johnny's is, there are some plant varieties for which they just don't have any heirloom varieties. This is the reason why I turned to Burpee's this year, to fill in some of those gaps.

Wesley Dios

There's another perhaps unintended link with permaculture in this approach, the idea of diversifying our systems as much as possible. Now, I readily admit that buying seeds from two sources hardly qualifies as radical diversification but it's a start. Plus, I've been too pressed for time to go through multiple seed catalogs and work out an order — just coordinating between two was about all I could handle. In the coming years, I plan to expand my ordering sources out to other companies — two that spring to mind immediately are Baker Creek Heirloom Seeds and Seed Savers Exchange. At the same time, I'll start saving seeds from my heirloom varieties to breed for specific characteristics, while also allowing many of the crops (such as lettuce and spinach) to become self-seeding annual groundcover within some of my permaculture forest gardens. That way, over the coming years, my seed sources will become a fractal of my overall permaculture systems — resilient, diversified polyculture designed to withstand several outside shocks, because it is a self-sustaining system.

Here's a list of the vegetable seeds that I still have in my inventory from last year:

- **Beans:** Fortex (pole fillet), Maxibel (bush fillet), Black Coco (dry), Scarlet Beauty (dry), Red Kidney (dry), Cannellini (dry)
- **Peas:** Sugar Snap
- **Cantaloupe:** Hale's Best Jumbo
- **Kale:** Red Russian
- **Broccoli:** Southern Comet Hybrid
- **Zucchini:** Black Beauty Hybrid
- **Eggplant:** Fairy Tale Hybrid
- **Cucumbers:** Marketmore 76 Hybrid
- **Carrots:** Imperator, Scarlet Nantes, Nantes Half-Long
- **Peppers:** Cayenne Large Red Thick (hot), Serrano Tampiqueno (hot), Jupiter Bell, Napoleon Sweet

Permaculture

- **Tomatoes:** Mortgage Lifter (beefsteak), Juliet (saladette hybrid), Cherokee Purple (beefsteak), Amish Paste (sauce), Red Baby Roma (plum/sauce), Gardener's Delight (cherry), Djena Lee's Golden Girl (yellow cherry)
- **Lettuce:** New Red Fire, Red Oakleaf, Romaine Cos Winter Density, Black Seeded Simpson
- **Spinach:** Tyee, Emu
- **Greens:** Escarole, Endive, Arugula
- **Herbs:** Cilantro, Titan Parsley, Rosemary, Stevia, Basil (Genovese), Oregano (Greek), Thyme (German Winter)
- **Potatoes:** Kennebec, Red Pontiac (seed potatoes taken from leftovers of current winter storage stock)

Here's the list of what I got from Johnny's Selected Seeds:

- **Radish:** Miyashige Daikon, Shunkyo Semi-Long Daikon
- **Tomatoes:** Juliet (hybrid saladette), Brandywine (beefsteak heirloom)
- **Sweet Potatoes:** Beauregard
- **Squash:** Waltham Butternut (heirloom), Black Forest Kabocha (heirloom), Raven Zucchini (hybrid)
- **Potatoes:** Dark Red Norland, Kennebec, Yukon Gold, French Fingerling
- **Onions:** Evergreen Hardy White, Bridger, Ruby Ring
- **Leeks:** King Richard
- **Lettuce:** Allstar Gourmet Mix, Encore Mix
- **Corn:** Nothstein Dent (meal), Red Beauty (pop), Double Standard (open pollinated sweet), Silver Queen (late hybrid)
- **Celery:** Tango
- **Broccoli:** Arcadia
- **Beans:** European Soldier (dry), Gita (yard-long), Red Noodle (yard-long)

Wesley Dios

Here are the seeds I ordered from Burpee's:

- **Carrots:** Danvers Half-Long (heirloom)
- **Corn:** Golden Bantam (yellow sweet heirloom)
- **Eggplant:** Black Beauty (heirloom)
- **Garlic:** Elephant
- **Shallots:** Holland Red
- **Pumpkin:** Rouge Vif d'Etampes (LARGE heirloom)
- **Spinach:** New Zealand (heirloom/perennial), Bloomsdale Long-Standing (heirloom)
- **Summer Squash:** Early Prolific Straightneck (heirloom)

The total of these orders came out to $237.90. If viewed through the lens of our consumer economy, this can seem like a lot of money. However, I view my seed order as not just investing in this year, but expanding my seed stock for years to come. Plus, when this expenditure is compared with how much money it would cost to purchase fresh, organic varieties of these vegetables from a farmer's market or Whole Foods, it really comes out as being quite minimal. Finally, by setting up a small "honor-system" farmstand at the end of our driveway for local passers-by, I can probably recoup a large portion of this outlay — and possibly even turn a small profit — just by selling the inevitable surplus from my gardens at a reasonable price.

CHAPTER 3- WASTE NOT, WANT NOT

One of the primary features of permaculture systems is that they produce more in yield than they receive in inputs and emit as "waste." Many of our systems in the modern, industrialized world seem as if they provide bountiful yields, but in most cases the yield is discounted when the inputs and "waste" are figured into the equation. For example, the United States "conventional" agricultural system is considered by many to be the most productive in the world. But when you factor in the amount of fuel required to run the tractors and combine, the amount of energy that goes into making the pesticides, herbicides and chemical fertilizer needed for large-scale monoculture operations, and the amount of topsoil and chemicals that are washed into waterways via erosion – the system actually provides a negative yield. The only thing that has kept it going so far has been the availability of cheap fossil fuels.

Wesley Dios

In case you're wondering, there's a reason why I placed the term "waste" in quotation marks. In permaculture, there is no such thing as "waste," because permaculture seeks to emulate nature, and in nature the concept of "waste" does not exist. Natural ecosystems seek to use everything, to effectively close the loop. Dead plants and animals provide fertilizer for new plants, which are in turn eaten by animals, and those animals are in turn eaten by other animals, which then die and are broken down by bacteria and fungus into a form that new plants can take up. The same can be said for animal defecation and urination. There's no such thing as "waste" in a truly sustainable system. This brings us to the heart of this chapter, and the rather simple example I can provide in how to close the loop and turn "waste" into a useful resource.

Every day, millions of people in New York City defecate and urinate in significant volumes of clean water, and flush that water into sewer pipes, through which it travels until it reaches the treatment Plant. There, immense amounts of energy are dedicated toward treating the sewage in this water with various mechanical, biological and chemical processes before it is safe to release into the waterways.

When most people look at these kinds of facilities, they likely see progress – sanitation brought to a city of close to ten million people. However, when I look at it, I can't help but be struck by the immense inefficiency of it all. Here we have a valuable biological resource that can provide massive amounts of nutrients for growing useful plants, and instead we mix it with perfectly clean water and use massive amounts of energy to then clean that water after the fact. While I'm not saying that we should immediately end sewage systems for all of New York City, at the same time you have to admit that there's a certain folly to it all.

Permaculture

I must admit that I'm somewhat guilty of this myself. We have flush toilets at my house, and although they don't go to a wastewater treatment plant, they do go to a septic tank that doesn't really provide any benefit toward closing the nutrient loop. However, since I have a pretty long commute to work; I have to relieve myself of my morning coffee every morning. Rather than stop at a gas station bathroom, I carry a used apple juice container with a tight, screw-on lid that I keep in my car. Part of the reason for this is that it's much faster than stopping at a gas station. But the other (and more important) part is that when I get home I take that urine and pour it either in my compost bins, or now, on dormant planting beds. Recently I started working my way along the berm that I built in my front yard, and I should be able to make it up and down that berm a couple of times before the planting season comes.

If you're second guessing my sanity here, let me tell you that there's some definite science at work here. Human urine is extremely high in nitrogen, phosphorus and potassium (the N-P-K of chemical fertilizer) in a form that is easily taken up by plants. It is also sterile when exiting the body, but serves as a great growing medium for bacteria. While many people dilute it and use the solution to water and fertilize plants directly, I prefer to use it to encourage the growth of bacteria and other beneficial microorganisms in my soil and compost. By doing so, I hope to encourage better long-term yields by helping to kick-start my soil-building process. I'm expanding this operation as well by keeping a bucket filled with wood mulch in the garage to be used as a urinal. I'll use the mulch, saturated with the nutrients from urine, to help fertilize crops and compost in the coming year.

I've also developed some basic concepts of how to better capture our human "waste" streams and turn them into productive resources. I could build an underground bio-digester to take the place of our septic tank, and turn our toilet waste into fuel for methane production, which can be used for cooking gas. A grey water diversion system would take water used for washing inside the house, and use it to recharge swales or to support a rain garden. Installing an enclosed lagoon where my septic leach field now sits could help produce vigorous growth in plants used for mulch and building materials. While all of these options are ones to consider moving forward, the important thing is that I'm doing something to help reduce my "waste" right now and thinking about ways that I can close the loop even more in the future. In doing so, I'm proving that we have abundance all around us that we can put to beneficial use.

CHAPTER 4- HOW TO SAVE WATER IN THE GARDEN

Freshwater is arguably the most precious resource on Earth, as demand increases every day but supply decreases steadily. In some areas, prolonged drought is forcing many to reconsider time-honored gardening practices. Even in areas with adequate rain, many choose to conserve water for environmental reasons, and some simply want to lower their water bill. Regardless of your motivations, there are many options available to help reduce water use from planting to harvest. Whether you grow vegetables or blossoms, consider these aspects of your garden to optimize it for water efficiency.

Soil Amendments

There are various amendments that can be added to the soil before planting to retain water. Sphagnum peat, vermiculite, and compost all add moisture retention, but they each come with other effects as well.

Sphagnum peat retains moisture very well and adds some friability, but is also nutritionally empty and very acidic. It's important to do a pH test on your soil before you add large amounts of peat, as an extremely acidic garden is not ideal for most plants. Peat is a semi-renewable resource harvested from bogs, and its sustainability depends on the rate at which it is harvested.

Vermiculite is a naturally-occurring mineral that is mined for its unique properties and used in a wide variety of commercial applications. Like peat, it contains no nutrients, but unlike peat vermiculite is pH-neutral. It can last many years depending on your conditions, and can be used in combination with any other amendment.

Compost is a great option if it's of good quality, however there are many inferior compost products on the market today and it can be difficult for a consumer to navigate the available choices. Many options contain "bio-solids," which is the trade-name for treated sewage sludge. While sludge is nutrient-rich and safe in terms of bacteria, it contains other traces of the human lifestyle such as heavy metals and pharmaceutical residue. Don't let these cautions deter you from using good compost, which retains moisture, adds precious humus to the soil, and is nutritionally ideal. You can always ensure quality by making your own.

Plant Choice

Some plants require more water than others. If you live in an area that doesn't get much rainfall, favor drought-tolerant plants that don't need too much pampering. Okra, garlic, peppers, sweet potatoes, and amaranth are all heat- and drought- tolerant vegetables, and there are many varieties of thirsty plants bred to thrive in dry conditions. For a true desert garden, try a delicious prickly-pear cactus!

In decorative beds, consider planting local wildflowers or drought-tolerant herbs, such as chamomile, lavender, sage, and rosemary instead of traditional flowers. They are beautiful and aromatic, they attract pollinators, and best of all they need just an occasional sip of water once they are well-established. The technical term for landscaping with low water use plants is "xeriscaping," and it's becoming a popular buzzword in the gardening community.

Garden Layout

Block-style planting, such as the square foot method, loses less water to evaporation than row-style planting, which exposes more of the soil to the open air. Group plants with similar water needs together so that you can manage water application by bed. Deep-rooted plants like tomatoes and squash need a lot of water early in the season, but can tolerate much less in the mid-late summer. Lettuces and sweet corn don't do well without constantly moist soil, but they can work together if you place lettuce on the east side of a bed and corn on the west. These thirsty plants can share wetter conditions, and a few hours of afternoon shade under the cornstalks will delay the lettuce bolting or becoming bitter.

Mulch

Adding a thick layer of mulch over the soil provides a protective barrier that absorbs moisture and keeps it from evaporating. Other benefits include smothering weeds, stabilizing the soil temperature, and preventing runoff erosion. Though wood mulch is a popular choice due to its tidy appearance, freshly produced wood chips can tie up nitrogen in the soil, leading to a deficiency in your plants. Aged wood mulches use less nitrogen, and bark chips even less. Some wood-based mulch, such as cypress, are environmentally unsustainable, so if you choose to use wood, be sure to do a little research before you make your purchase.

Fallen leaves and dry grass clippings are an inexpensive and somewhat more natural option, both of which add nitrogen to the soil as they decay. At the end of the season, they can be turned under the soil to slowly compost and add humus. You must make sure that grass clippings are dry before you use them as mulch. Fresh clippings are so nitrogen-rich that they will "burn" many plants as they break down.

Irrigation

Traditional watering by hose can lose water through soil saturation, which results in runoff and evaporation. Drip irrigation avoids this issue by watering slowly over a longer period. (Consider the effects of a torrential afternoon thunderstorm compared to a prolonged, gentle shower.) Drip irrigation is also applied close to the soil rather than being sent through the air by sprinkler or hose, further reducing the amount lost due to evaporation.

Drip systems come in many forms, and can be as complex as a professionally-installed timed drip hose or as simple as a few half-buried milk jugs with holes poked in the bottom. Soaker hoses are a good mid-spectrum option, touting simple installation and relatively low cost.

Rain Collection

Harvesting rainfall is perhaps the simplest way to reduce groundwater use. The easiest way to implement collection is to install a barrel under the downspout of your gutter. Remember to keep the vessel covered; otherwise it will quickly become a breeding ground for mosquitos. A spigot installed at the bottom of the barrel will make it easier to fill your watering can, but

Permaculture

remember to elevate the barrel in this case (perhaps with cinder blocks), so your can will fit under the spout. Even in areas with very little rainfall, a barrel can collect those few summer showers and provide several gallons of water to sustain your garden. Some dry areas suffer from alkali groundwater, and most plants will prefer pH-neutral rainwater when it is available. Be sure to check your local regulations before collecting rainwater; as in some states it requires a permit.

Every garden is different, so think about what alterations will be the most efficient in your yard. If you incorporate just a few of these tips into your garden this year, you can become a better neighbor to those downstream and lower your water bill at the same time!

CHAPTER 5- PERMACULTURE ETHICS- UNDERSTANDING GIFTS

A short time back, we received a Christmas card from my wife's grandmother. When I said something to my wife about getting the card, she immediately said that she was upset because her grandmother always sends a check for holidays and birthdays, and seeing how she is on a very limited and fixed income, she cannot really afford to send anything. More recently, we were getting a few Christmas gifts ready for my parents in anticipation of them coming over for an early holiday gathering. My wife expressed concern that we were not giving my parents enough in return for the gift they gave us and told me that she felt bad about this.

Permaculture

While feelings of displeasure surrounding the receipt or giving of gifts may seem contradictory on the surface, the reality is that this is the reaction that many of us have more often than we care to admit — myself definitely included. It also shows the way that gift-giving has been almost completely marginalized within our society, for us to feel a sense of genuine unease over something that should bring us happiness instead. I think that the source for this unease is something that very few people recognize — the way that the rules of market exchange have gradually overtaken many areas of our lives that they have no business entering, with the exchange of gifts being a major instance of that intrusion.

In order to explore this issue, it's important for us to look at the basic mechanics of gift economies versus market economies, because they are decidedly different from one another. The market relies primarily on immediate, quid-pro-quo exchanges between people who often have no other relationship than the moment of that exchange of money for an item or service. Once that immediate exchange is satisfied between purchaser and vendor, there is no outstanding obligation between the parties. Contrast this with gift economies, which not only rely upon continuing obligations to function properly, but exist within broad webs where those obligations are not necessarily paid back to the party who gave the initial gift, but are more often paid forward to others.

When looked at in this manner, I think that it is understandable how we have such issues with gifts when we are conditioned countless times each day to think in terms of market exchanges. To take it a step further (and perhaps make it more controversial), I think that greater engagement of the gift economy has the potential to undermine the market economy's hold on so many aspects of our daily lives — and enable us to live more aligned with the permaculture ethics, particularly the third ethic of return of surplus.

Charles Eisenstein has written extensively on the nature of gift economies in his book, "Sacred Economics". I don't want to go too deeply into his writing in this piece, but one of the major themes he takes up is the subject of gift economies, and the central role that they played in our daily lives for the overwhelming majority of human existence on this planet. Eisenstein likens our current market-dominated economic landscape to a monoculture of one single plant (such as corn) spreading across thousands and thousands of acres. This kind of arrangement can be held together through massive interventions — at least for a time — but overall it is fragile and fleeting, and good neither for the corn nor for the end user. I don't consider it a stretch to look at the goings-on of the financial sector over the past few years and conclude that what the money economy is currently doing is not good for most of the citizenry in the "developed" world, or for the long-term viability of the currencies of those nations.

Another theme that Eisenstein visits in his writing and talks is the way that an increased reliance on money creates a sense of scarcity — and although it seems contradictory, the more money that people have, the tighter they hold on to it for themselves. This can be supported by the statistic that people from lower income brackets tend to give a higher percentage of their income as compared to people from higher brackets. Simply put, the more money a person or family takes in, the more they use that money to pay for a broader range of goods and services (thus increasing their reliance on the financial system), and the less likely they become to readily share what they have. I have seen this dynamic at work in my own life. As my household income has gone up, I have actually found myself becoming more preoccupied with money than I was before, and consistently finding excuses to put off giving — even as my surplus has grown. It is only through reading, listening and thinking about the nature of gift economies that I am even able to recognize this phenomenon at work and come up with a plan to counter it.

The first step I have taken to better orienting my financial life with the permaculture ethics through gift giving is to change the way that I look at gifts given to me as well as how I give gifts to others. Although I still have the urge to protest gifts when offered, I force myself to open up and accept them, with all of the gratitude that acceptance entails. When it comes to giving gifts to others, I have expanded my definition of a "gift" — it could be something material, but it could also be sharing my time with someone or providing a skill that I have. In the event that I feel like someone else has given me more than I have given them, I do not allow myself to be constrained by the need to pay them back, but rather open up the possibilities to paying their gift forward in the future to someone in need.

Taking this further, my wife and I are planning out scheduled giving as a part of our regular budget — no different than food, utilities, or savings. By doing so, we are forcing ourselves to regularly engage in the gift economy. My goal is for this to eventually help us to displace our over-reliance on the monetary/financial economy and increase our household resiliency in the process. For an example of how far you can take such an engagement of the gift economy — and the rich life you can discover as a result — I recommend checking out "Radical Possibilities with Ethan Hughes" from The Permaculture Podcast with Scott Mann.

CHAPTER 6- PERMACULTURE DESIGN- NONVIOLENT COMMUNICATION AND NATURE

As I said in Chapter 1, Permaculture is often described as a revolution disguised as organic gardening. Permaculture Design is a term coined by Bill Mollison referring to the discovery and application of nature's perfect design through observation and appreciation of the multitude of inherent qualities and relationships replicable by humans who are oriented and dedicated to caring for the earth and people to create sustainable environments for living in health and abundance. A way of designing systems and, strategies to support those systems by clearly identifying the whole environment; it's many participants and unique qualities, resources/energy and relational functions. It recognizes choice of action in honor of them as the foundation for encouragement and succession toward sustainability. Permaculture Design is a commitment to nurture the whole by way of honoring it's parts. As applied to land-based and invisible structures it is an exploration, discovery and appropriate use of nature's patterns and diversity with specific attention to meeting needs most efficiently and maintaining focused attention on enhancing the quality of life and establishing the environment for sustainable living.

Communication/An Invisible Structure: a presentation bridging our awareness of land-based systems and communication as an integral, natural system worthy of Permaculture Design attention toward the transformation of awareness. Exploration of models for influencing environments toward sustainability through language and communication facilitation will include Nonviolent Communication and Dynamic Facilitation.

Permaculture

Permaculture Design Collaboratives

Let your friends, family and neighbors be involved in the design and implementation of the transformation of your yard, lawn, walkway, porch or other aspect of your personal habitat. With my skills in Dynamic Facilitation, Nonviolent Communication and Permaculture Design, I guide small groups in arriving at decisions based on the protocol of Permaculture Design. You and those who choose to participate in your project will be an integral part of observing the site, identifying elements and functions, resources, and designing to meet the "clients" needs and serve all the relationships of the environment.

Ever wonder why he/she just said that? And, how can I possibly get him/her to understand? Or, do you want some skills to prevent miscommunication, unnecessary anger, doubt and worry? Sure, everything is fine. Till it's not and you're stuck with the same way of communication and habits passed on for centuries. That stuff doesn't just go away. You can unlearn the teachings of war---the blame, shame, guilt and judgment that keep us disconnected from life and the joy of contributing to each other's well-being.

Any demands that are made (or heard) disconnect us from our desire and availability to meet each other's needs. How is it that our love can be communicated in a way to bring out anger, fear, shame and doubt? When I'm deeply concerned, how can I express it compassionately? When I'm scared and need the well-being of my child, my friend, myself how can I make a clear request from the place within me that knows love and finds courage? Nonviolent (Compassionate) Communication is a process for transforming situations into understandable opportunities for collaboration. It is a practice of focused attention with the intent of creating the connection wherein we are more likely to give naturally and meet needs for peace while creating fulfilling relationships.

Nonviolent Communication/Compassionate Communication

Nonviolent Communication/Compassionate Communication is a model and tool for practice; a personal process wherein we can become more alive, in harmony and compassionately available to meet our needs. You are invited to experience and cultivate a skill for engagement in authentic communication:

Honesty: The courage to acknowledge true feelings and accept these messengers. Honesty can lead us to identify our needs and take specific action to fulfill them. I want to be discovered and embraced. I can begin personal and social transformation by listening. A person's need for connection and empathy, safety and freedom can be met as anger, guilt, fear, anxiety are expressed.

Permaculture

Self-Empathy: When no one else is able to listen to you, you can focus on your experience in a way that creates life-connection within you. This allows for peace, time and space so that you are available again, to be in authentic communication with others, finding compassion is alive.

Empathic Listening: Imagine some time and space to tell your story, share your beliefs, values, visions, hopes and fears. To feel and remember. To question and hunger. A time and space where there is no "history" between us, no agenda, anxiety or fear of creating more anger, distrust or annoyance. To speak of your day, all your relations and be heard for all you are worth. A place in time where we can go anywhere together. We've never heard it before. It's your expression in this moment.

Celebration: We will know when we have arrived at the place of life affirmation. Here, I no longer seek approval, praise or punishment as motivators to contribute to our experience. I am alive. NVC/Compassionate Communication is a model for transformation of life-alienation toward life-connection.

Mediation/Conflict Resolution

Nonviolence communication courses offer skillful reflective listening, specific communication guidance and otherwise create a safe, comfortable environment to feel and express authentically. Participants are likely to come to mutual understanding, compassionately consider solutions and create agreeable solutions.

- Friends, family, small group
- Teen, parent
- Marriage/divorce/reconciliation
- Newly wed, new parents
- Rehabilitation, extended family resolution

Wesley Dios
- Land rights, business partnerships
- Neighborhood disputes, board meetings
- Musical/art groups

Facilitation

This service is one on one so that you are heard and revitalized. This service meets needs for empathy, connection, understanding, compassion. It is likely that during a facilitation session you will gain clarity, resolve internal disputes and determine solutions.

Training and Practice

You are likely to understand the tool as a practical means to authentic communication in harmony and power with each other and will have experiences working with the model in various exercises and live communication for small and large groups applicable toward:

- Service Development Training
- Schools
- Conferences
- Families
- Community/Organization/Partnership Retreats
- Health/Healing Retreats

From the models' basic, clear- cut form we can see how tangled we get by way of the communication habits we have learned and live by. Participants can begin to hear patterns that go almost unnoticed and, accepted by humanity that has been taught for thousands of years to judge, advise and battle with intellect and engage in emotional terrorism. You will be guided in practical examples of listening, responding and creating relations from a

Permaculture

place of human understanding. How is it that we are motivated in any given moment to take action?

Practice Groups: The practice is all about you. It is about self-care, and responsibility to others. We come together as mutual participants in learning and gifting of ourselves. Each practice group is a two hour guided, interactive experience in using the NVC (nonviolent communication) model offered out of reverence to nurture as we share personal experiences "past and present", role play, discuss our habitual ways of communication for understanding and motivation toward this conscious process of focused attention, self-expression and cultivation of ease in making connecting requests.

Workshops: are intensives for groups designed for the experienced or new-comer to further understanding and use of the NVC model in daily life. Participation in a workshop allows for heightened awareness and recognition of what is possible in the practice of Nonviolent (Compassionate) Communication. It is likely you will develop a new capacity for honest communication and empathic connection as agents for transformation of relationships.

Service and Relations Development Training Program: the business/service organization will be considered as a complete environment, all functioning aspects affect each NVC training program design. Programs include ongoing NVC training for employees and volunteers bringing into place a foundation for focused, practical means to cultivate communication skills for organizational development, as continuing education for staff and enhancement of customer/partner relations. Complements existing programs and inspires an authentic, creative approach to networking and resource development while carrying out core missions.

Individual training sessions: privacy, peace, trust, freedom, personal growth. Beginning with a series of six sessions, as a means of developing empathic listening skills and courage toward honest communication. To fortify family and workplace relations and, inter-professional communications and orientation. Examples of professional service development roles to enhance: hospice, teachers, counselor, social worker, foster care, human relations development coordinators, organizational, directorship, environmental/architectural design, social activism.

Family sessions: There is no environment like family for personal challenge and vibrant discovery. Together, we commit to a scheduled series of guided, inner/inter-personal reflection and authentic communication, developing skills in living honest, authentic relations.

Toddlers Tracks

My extensive experience as a professional childcare provider in many settings and my adventures walking in the neighborhood with many a toddler has prepared me to offer this with confidence. From the practice of Nonviolent Communication, which I've been immersed in for 9 years and teaching for 7, I've gained an ability to connect and communicate with infants and young children to meet needs for safety, freedom, peace and autonomy. Too many times, I've walked off playgrounds because I don't want to hear any more of the communication that is offered from adult to child. So much instruction, insistence, interjection! I wish they could relax! I've concluded that many adults just don't know how to relax and enjoy observing, encouraging and or playing with the child.

Training for parents to let little Johnny listen to himself and the rest of the world. I'm okay; leave me alone, I'll figure it out! My hope is for the parents to remember the joy of exploring, innocence and inquiry that calls them to engage in adventure and play. To trust their child and cultivate deeper listening skills to support meeting their basic needs. Say less, don't think you know it all, and come along for the ride. One parent is invited to come with the four of us. For one month, we alternate the parent participation. Once a week during that same month, I offer the same group of parents to come together and practice compassionate communication. To hash out their worries, frustrations, regrets and disputes and see clearly and beyond what they were able to do in that moment and remember another day will offer a million more opportunities to grow, create and love.

Nature is our common ground- Advanced group collaborations

The third part of this adventure, if you so choose to take it, brings the families together to design each other's outdoor living spaces. Want to transform your front or backyard? Porch, balcony or walkway? Using my skills in Dynamic Facilitation, Nonviolent Communication and Permaculture Design I often facilitate the creative collaborative between small groups of people who want the experience of hands-on design. An aspect of every design will be the ability to implement the design having identified resources available. The process includes harvesting information from participants about surplus (you know all that stuff in the garage, at your cousins' place, down the road, in the alley...) and putting it to use. This could be small renovations, adding a garden, play structure, or basic, efficient water harvesting system by moving some earth around to "slow it, spread it, and sink it". We could go all out! We will arrive at decisions by applying Permaculture Design. I could do this myself but, I think we'd have more fun doing it together and, we all learn! Also, as a team, it's much more cost effective.

Permaculture Design Collaboratives-

Wesley Dios

Let your friends, family and neighbors be involved in the design and implementation for the transformation of your yard, lawn, walkway, porch or other aspect of your personal habitat. With my skills in Dynamic Facilitation, Nonviolent Communication and Permaculture Design I often guide small groups in arriving at decisions based on the protocol of Permaculture Design. You and those who choose to participate in your project will be an integral part of observing the site, identifying elements and functions, resources, and designing to meet the "clients" needs and serve all the relationships of the environment.

Dynamic Facilitation

This service provides extensive support for any group or organization to realize the creative collaborative as they come together to address a problem with the intention of arriving at a solution authentically. For any business that is breaking free from the lie of "consensus" and the trauma of participating in meetings that end up in a division of the group or otherwise concluding while in disharmony, disconnect and having arrived dishonestly.

I'm motivated to reciprocate as a vehicle for support as I've learned so much and I am certain that the change from a society caught up in a domination system can happen if we are once again tuned-in to /awakened to listen deeply, communicate honestly and design our world sustainably. Every conscious action makes a difference. Learning to follow intuition, inner guidance is a step to discovery of what the world has to offer each of us. I'm motivated to facilitate healing and encourage a cultural revolution. I want to inspire trust in one's self and awaken dedicated passion to actions that include the art of invitation. Inviting people to share their stories, to ask for what they need and be available to each other for the unexpected exchange of purposeful resource. Our language and fear have kept us from making these connections---there has been

Permaculture

a movement, cultivated by fear and domination, away from intimate exchanges wherein people come together face to face, co-creating neighborhoods, communities, services, and businesses. Our fears can indicate to us unmet needs instead of owning us. When we honor feelings as messengers and notice the invitation to choose then we can embark on a discovery of our full potential. Language can transform our living, out of habits and programming that we've embodied since it was first heard. As we are human and living of this planet, the relationship is undeniable. The natural resources thrive as humans care to cultivate understanding and appreciation and, take reciprocal action. I'm motivated to help all who seek to create trusting relationships with self and the natural world. To relieve them of the pressures of the consumer world, fuel their passion and give them skills to successfully use that energy in their lives.

ABOUT THE AUTHOR

Wesley Dios majored in agrarian studies at university but was intrigued by permaculture. It was something that he had heard about and was determined to find out all he could about it when he had the time. Upon the completion of his degree he started that research and even enrolled for a course on campus.

Wesley loved the environment and was determined to do anything that he could to make more persons aware of what they could do to sustain it especially the farming community. His studies even led to him going to the place where permaculture originated, Australia.